Growing and Showing
Vegetables

Tom Fenton

David & Charles
Newton Abbot   London   North Pomfret (Vt)

British Library Cataloguing in Publication Data

Fenton, Tom
    Growing and showing vegetables. – (Growing
and showing series)
    1. Vegetables
    I. Title   II. Series
    635   SB320.9

    ISBN 0–7153–8577–1

Photoset in Souvenir by
Northern Phototypesetting Co, Bolton
and printed in Great Britain by
Redwood Burn Ltd, Trowbridge, Wilts
for David & Charles (Publishers) Limited
Brunel House   Newton Abbot   Devon

Published in the United States of America
by David & Charles Inc
North Pomfret   Vermont 05053   USA

Contents

Introduction

The purpose of this book is to help exhibitors and intending exhibitors of vegetables to improve on the growing, preparation for showing and final presentation of their subjects on the showbench. One occasionally sees quite good specimens unplaced because of the lack of finesse prior to staging. It is important that great attention to detail is given to every aspect from the very beginning, right through cultivation to the vegetable's final position on the showbench.

The dates mentioned in this book are those best for growers in the north of England; adjustments must be made for Scotland or areas in the south of England. In any case, make contact with your local District Association of the National Vegetable Society.

I would like to thank my daughter Mrs E Winwood, and Mr Jack Wood, for their valuable assistance in the production of this book.

1 General Practices

Crop Rotation

While it is necessary to have a certain amount of crop rotation, some crops may be grown successfully year after year on the same piece of ground. These include onions, leeks, runner beans, etc. Quite often there is some particular reason for doing this, such as growing runner beans on a site sheltered from strong winds, or onions and leeks on a nicely sunny site.

There are, however, certain rules which must be kept. The remaining area of the vegetable plot should be divided roughly into three sections, for a three-year rotation:

Area 1: peas, beans, potatoes, celery.

Area 2: carrots, beetroots, parsnips, swedes.

Area 3: cabbages, cauliflowers, brussels sprouts, savoys.

The following year, move the peas, etc to Area 2, the carrots, etc to Area 3 and the brassicas to Area 1. Complete the rotation in the third year.

The importance of rotation is to minimise the risk of carrying over pests and disease. Also, brassicas are hungry crops, requiring liberal amounts of farmyard manure, while root crops need only artificial fertilisers, with plenty of phosphates and potash.

Preparation of Sites

It should be the aim of every gardener to complete all digging by the end of October if possible, as it is so much easier to do while the soil is still warm. The frosts can then break down the soil during the winter and bring it to a nice friable state for the spring work.

Hygiene

Anyone wishing to grow top-quality vegetables should be careful to observe general rules of cleanliness in the garden. All vegetable material which has served its usefulness should be removed and consigned to the compost heap or burned. Leaving decaying material lying about can only lead to further contamination. Greenhouses and plant containers, such as pots and boxes, should be washed before being used again. Use soap and water, and a good sterilising agent to make sure there is no carry over of infection from one season to the next. If there is any suspicion of pests or disease in the border soil it too should be sterilised, using a reliable soil steriliser.

Composting

Waste vegetable matter can be converted back into a valuable form of organic material, highly fertile in plant nutrients. Use a container that will permit air to reach the compost, and give it a cover to keep out the rain – if it gets too wet the conditions for breaking down the vegetables are less good. An activator may be added, as the heap accumulates, to build up the bacteria which break down the raw material into humus. A quicker but more laborious way is to turn the heap every four weeks or so at the end of the season, applying an activator until the breakdown is completed. This can raise the temperature of the compost, if farmyard manure is incorporated, to 160°F (70°C), at which point most harmful competitors are killed.

Manures and Fertilisers

Both of these are beneficial to and essential for plant growth. Manures, such as farmyard manure, hoof and horn, dried seaweed, etc, are organic. Fertilisers, such as sulphate of ammonia, superphosphate, sulphate of potash and their various combinations, are chemical. Both are very important in producing high-grade crops; they should be used carefully and adjusted in order to produce the desired type of growth.

Remember that, generally, nitrogen is responsible for leaf and stem development, phosphates for production of a good root system, and potash for good health, maturity and resistance to pests and disease. Lime is usually needed only to ensure that the soil remains in a slightly acid state: ie with a pH of 6.3–6.5. Trace elements are usually supplied in sufficient amounts by applications of farmyard manure in the bed or border, but they can be added in chemical form to composts.

Choice of Varieties

When selecting varieties of vegetables for exhibition, it is vital to set off on the right foot. There are several seedhouses today that can supply suitable seeds or plants. However, many of the more successful exhibitors have their own strains of various subjects; some of them may be quite prepared to sell plants to other exhibitors. A common method of acquiring good stock is by exchanging with another exhibitor, if you have something that he particularly fancies.

Successful Exhibiting

The old saying 'The more you know, the more you know you don't know' is certainly true of a successful exhibitor. You must never be satisfied that you know it all, because it is certainly possible to learn something from a novice. Once you decide to try your luck at the larger shows, you will find that the arts and crafts of showing vegetables are operated to a high degree, and the standards are so much higher than local shows. This is where your education in showing really starts: you will learn from your own mistakes, and also those of others (conversation with fellow exhibitors is one of the joys of the show scene). To become successful at national level you need attention to detail, an enquiring mind, and quite a slice of luck.

A good exhibitor will accept defeat with a smile, and a determination to do better next time; and will always be ready to offer advice to novice exhibitors. The decision of the judges must be accepted, even if you do not agree.

The standard of vegetables shown today has improved quite remarkably from that of even twenty years ago. Professional plant breeders have contributed to this in no small measure, but good growers and exhibitors have also played a large part. Showmen have benefitted by picking the brains of plant breeders and applying this knowledge in their own areas to produce winning specimens.

In the following discussions on the various vegetables, I often mention the categories of form, condition, size and uniformity which are used to allocate points. Below is a brief explanation of each category.

'Form' is only applicable to those vegetables for which a certain shape or type is required, eg dwarf beans, runner beans, carrots, cucumbers, onions, parsnips and tomatoes. Under 'condition', judges look for cleanliness, freshness, tenderness, and absence of coarseness or blemishes. A large size is only an advantage if accompanied by quality. Large specimens of good quality are much more difficult to grow than smaller specimens. 'Uniformity' means that all the specimens in a group of the same vegetable should be alike in size, form and colour.

Staging a Collection

Much depends on the number of different vegetables required by the schedule. The small collections of 3 or 4 types are usually staged on the flat bench within a limited area. The collections of 6 or 8 are most demanding – to win in these classes at top shows the standard of every vegetable must be very high. The schedule usually states 'A collection of 6 (or 8) kinds of vegetable (one dish of each). A 'dish' is one cultivar of one vegetable, displayed on a stand (see page 33).

It is essential to choose those vegetables which are allocated high points. For a collection of 6, a suitable selection would be: 6 sticks of celery; 6 cauliflowers; 6 onions; 10 carrots; 12, 18 or 24 peas, and 6 potatoes. For a collection of 8, choose also from: 6 or 10 leeks; 6 parsnips; 12 or 18 runner beans, and 6 or 12 tomatoes. Show preparation is the same as for individual classes.

Tie the celery and leeks in pyramid form, 3, 2 and 1. Remove all leaves from the cauliflowers and fasten them to the board (by nails driven through from the back) in a triangle – 3 at the bottom, then 2 and 1 at the top. Garnish the cauliflowers with moss-curled parsley. Stage the onions in the centre, 3 at the back, then 2, and 1 at the front. Stage carrots and parsnips 3, 2, 1 on either side of the onions, with the whip facing to the front and slightly inwards. Stage peas in a wheel shape or just a straight row. Display tomatoes or potatoes on plates.

A good colour balance is also important, so place contrasting colours together, eg, carrots with parsnips, pink celery with blanched leeks, white potatoes with red tomatoes, and green peas with straw-coloured onions.

Cover with paper to prevent discolouring, and remove it just before judging starts. Then spray everything (except the onions) with clean water to present a glistening appearance.

2 Common Vegetables Grown for Exhibition

Aubergines

This is a relative of the tomato and should be sown about the middle of March to ensure sufficient fruits to show through the season. It is not widely shown individually but is more likely to be found on large displays.

Sow seeds two or three per 3in (7.5cm) pot, in a good seed compost and keep at a temperature of 60°F (15°C). The weakest seedlings in each pot should be removed, leaving the strongest one to be potted on into 5in (13cm) pots, and to be planted eventually in the greenhouse border at a spacing of 18in (45cm). Being a fairly strong grower, the aubergine requires plenty of water and nutrients at the root, but the atmosphere should be on the dry side. Remove the earlier fruits while they are small, to enable the plant to develop and build up sufficient energy to produce large, solid, shapely, well-coloured fruits, free from blemishes.

Points awarded are: condition 9, size 4, uniformity 5, giving a total of 18. Defective qualities are misshapen, hollow, shrivelled or poorly coloured fruits. Each fruit should have the calyx attached with approximately 1in (25mm) of fruit stalk. The exhibit should be staged on a white paper plate or 'as according to schedule'.

The variety Moneymaker is suitable for exhibition.

Beans, Broad

Sow seeds at fortnightly intervals, from early April until late May, to cover a show season; or, alternatively, raise in 3in (7.5cm) pots and plant out 6in (15cm) apart. Allow them to develop only one stem per plant and stop these when about six clusters of flowers have been produced. At each cluster the beans must be restricted to two pods.

Although you should spray with an appropriate insecticide against blackfly, the above method of culture will, by its very nature,

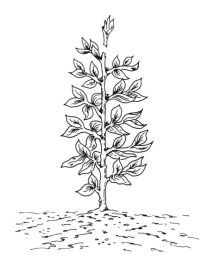

Top removed from a broad
bean plant to minimise blackfly

minimise attacks from this pest.

Support the plants against strong winds to keep them in an upright position, and make sure that the pods can hang freely, as it is important that they develop straight. When the pods are picked, they should have a short length of stem attached; and those finally selected should be large, fresh and well filled, with clean skins and tender seeds. Wrap them in a damp cloth to retain their freshness and keep them in a cool place. Staging is usually carried out on the flat bench, with the pods placed side by side and all pointing the same way. Shorten the stem to show a fresh cut.

The points awarded are: condition 6, size 5, and uniformity 4, making a total of 15.

Exhibition Longpod is an excellent white-seeded variety, capable of producing eight or nine seeds per pod.

Beans, Dwarf

These also must be sown in succession to produce ideal specimens suitable for exhibition. The best pods are produced in a greenhouse or under cloches, approximately nine to ten weeks after sowing. The most suitable method is to sow two beans per 3in (7·5cm) pot, in a loamless compost. Sow about fifteen to eighteen pots at each sowing, at intervals of ten days. When the seedling's seed leaves are fully developed, the weaker one should be cut off at soil level.

10

The dwarf bean likes to be planted on ground that has been well manured after the previous crop, with an additional 4oz (120g) of a base fertiliser with a high potash content. Put the plants out into the border, 12in (30cm) apart each way, with a 2ft (0.6m) cane pushed in by each plant. Only the centre shoot should be retained; remove all laterals. This centre shoot will produce four or five trusses of flowers. As soon as the small pods are developing, restrict them to about six to eight per plant, according to its vigour. They develop rapidly under glass, producing pods 9in to 10in (22cm to 25cm) or more in length.

Selection
It is necessary to measure the pods daily when they are nearing maturity because they must be cut off with about 1in (2.5cm) of stalk attached. Decide which length of pod to aim for, then cut each morning as they attain the desired length. Wrap them immediately in a damp cloth, or stand them, stalk end first, in a jar of clean water, and store in a dark, cool shed. Picking may start seven days before the show.

Staging
Trim the stalks to show a fresh cut. Lay them flat on the bench, on black velvet, with the stalks all pointing the same way, leaving a small space between each bean. An outstanding exhibit will have straight, tender, fresh pods, 10in (25cm) or so in length. The points awarded are: condition 6, size 3, form 3 and uniformity 3, totalling 15.

Suitable varieties are The Prince and Masterpiece.

Beans, Runner

You need at least two sowings of runner beans to cover the season, but, as germination is usually very good, it is only necessary to sow one seed per 4in (10cm) pot, in a loamless compost. Sow under glass, first in early May, and then ten days or so later, allowing ten per cent extra at each sowing. This method is sure to give you a full row at planting time, which is early June.

Preparation of border
Runner beans require a good sunny border, well sheltered and rich in humus. As with all legumes, they produce their own nitrogen

requirements. An application of superphosphate at 1½oz per sq yd (45g per sq m) and sulphate of potash at 2oz per sq yd (60g per sq m) should be incorporated during April, followed by a dressing of lime (according to a pH test) at planting time.

Erection of framework
You need to erect a framework before planting time to support the beans. An ideal way of growing exhibition beans is to plant a double row 30in (75cm) apart, with at least 15in (37cm) between the plants in the row. So, erect stout poles at 8ft (2.4m) intervals between the rows, with a double wire stretched along from pole to pole at a height of just over 6ft (2m), and push the canes from both rows (alternately) between the wires at the top, twisting the wires between each (see diagram). This will provide a very rigid framework against strong winds.

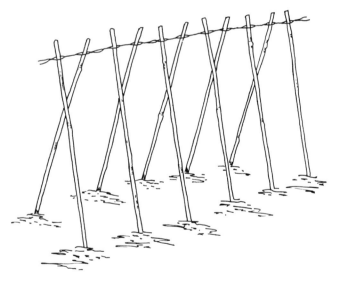

Framework of canes and wire
to support runner beans

Routine culture
As the plants grow, sweet-pea rings will keep the tips in close contact with the canes, to induce them to twine themselves round instead of wandering. Remember that the runner bean always twines anticlockwise! When the bines have reached the desired

Runner beans climbing anticlockwise round canes. The bottom trusses have been removed

height they must be stopped, allowing two leaves beyond the last flower truss. Remove all laterals as they develop, to direct maximum energy into the production of pods. It is also an advantage to remove all flower trusses up to a height of 2½ft (75cm) so that a large plant, with plenty of foliage, is produced before the demand for lush pods takes place. Not more than two of the straightest beans per truss should be retained.

Selection

As with dwarf beans, you must measure and cut each morning, starting about one week before the show. To retain freshness, wrap in a damp cloth, ensuring that each bean is kept straight. Alternatively, roll a lath, slightly longer than the beans, into black polythene, then add pods as the rolling continues. Store until show day in a dark cool place.

Staging

Trim off the end of the stalk and place the beans on a piece of black velvet to enhance their appearance. They must be spaced slightly apart with the stalks all pointing the same way. The merits of runner beans are the same as for dwarf beans, but the allocation of points differs: condition 8, size 3, form 3 and uniformity 4, totalling 18.

Good varieties are Enorma and Achievement.

Beetroot, Globe

This vegetable is most accommodating, as any soil (except chalk) will produce nice roots, providing that it was manured for the previous crop. Freshly manured land will produce fanged roots, which of course are useless for exhibition. April is early enough to start sowing – every three weeks until July. Short rows with three or four seeds per station, 6in (15cm) apart in the row, will give a good succession of nice young roots right through to the late November shows. Seedlings must be singled when large enough to handle.

Selection

Roots the size of a tennis ball are ideal, as larger ones will tend to lose colour, and colour is very important. Wash them down with a sponge (brushing can easily bruise the skin). Remove any fine rootlets, but leave the tap root intact. Some exhibitors insist on suspending the roots in water, with approximately 1oz (30g) of

Retain the single tap root when preparing globe beetroot

ordinary table salt per gallon of water, to improve the colour. The benefit of this is debatable, as soils in the same plot or garden can vary tremendously, producing well-coloured beetroots at one end, but poor specimens at the other.

Staging

Roots are usually staged on plates or just on the flat bench, and sprayed with water to keep them fresh. They may be tied to form a pyramid, or just laid down to form a triangle. Leaves should be removed, leaving not more than 3in (7.5cm) of leaf stalk. The judge will cut the roots to examine the colour, as more points are awarded for this than for other categories. Points awarded are: condition 5, colour of flesh 6, uniformity 4, making a total of 15.

Suitable varieties are Boltardy and Crimson Globe.

Beetroot, Long

These are grown, like carrots, in large containers, filled mainly with sand, or a similar material with little nutrient value. Bore sowing stations in the sand, approximately 3ft (90cm) deep, and fill them with a compost similar to John Innes No 2, which has been passed through a ¼in (6mm) sieve to remove any stones or other materials, which might bruise the skin. Sow three or four seeds at each station, and thin out when the seedlings are large enough to handle, leaving the most promising specimen at each point.

Selection
Before starting to pull, check round the crowns to match up for size and avoid wasting roots which may be required for a later show. Wash the beet immediately after pulling, removing all tiny rootlets down the long tap root, and leaving as much 'whip' as possible – long beet should taper down gradually to the thread-like extremity of the root. Wrap each specimen separately in damp material to prevent rubbing, and pack in boxes for transporting.

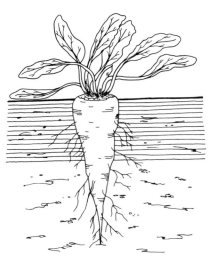

Long beetroot should taper down gradually to a thread-like root. Remove all the fine rootlets when washing long and globe beetroot

This type of beet is of little use except for exhibition, so it is not listed in many catalogues. The best way for a beginner to obtain some seed is to talk nicely to an exhibitor, who will no doubt be pleased to assist a fellow competitor.

Staging
This is usually done on the flat bench, with the specimens arranged as a fan. The leaves should be removed, leaving 3in (7.5cm) of leaf stalk. Points are awarded as for globe beetroot. Spray with water to maintain freshness.

Good exhibition varieties are The Vale and Long Black.

Cabbages

In the world of exhibiting, cabbages do not merit many points. Plants may be obtained (from a nurseryman) ready for planting out in May. To produce good hearts, you must give them plenty of space (2ft or 60cm apart each way), and make sure they are well firmed on planting.

Routine spraying with a good insecticide is necessary to avoid damage by caterpillars, etc, and, as brassicas are ravenous feeders the ground must have a liberal application of well-rotted farmyard manure in the previous autumn. Just before planting, a good application of lime is usually required to bring the pH to neutral or slightly alkaline, ie pH 7–8. This will also reduce the chance of infection from club root.

Selection
The size of the specimens is determined by the variety you intend to show. They must be well shaped and fresh, with solid hearts and good surrounding leaves. When selecting, do not press on the heart, as this will remove the bloom. Hold with both hands under the heart leaves which have opened, and squeeze gently to assess firmness.

Staging
Display the heart by turning back the outer leaves slightly. The show schedule will tell you whether the roots must be attached or cut off.

Suitable varieties are Minicole (small with oval heart), Derby Day (larger with round heart), and Hispi (with pointed heart).

Carrots, Long

Good specimens of this vegetable are not easy to grow, so the usual practice is to grow them in well-raised beds, barrels or large drainpipes containing a sandy mixture low in nutrients. To produce those long tapering roots, as seen at the top shows, bore holes to a depth of $3\frac{1}{2}$ft (105cm) and fill them up with a compost which has been passed through a $\frac{1}{4}$in (6mm) sieve. The compost must be of an open nature (to drain well), with an additional $\frac{1}{2}$lb of a well-balanced fertiliser per bushel (225g per 36 litres).

To produce specimens for early shows, sow in early April, putting in several seeds per station. It is most important that some

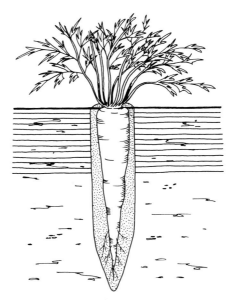

Long carrot growing in a hole filled with finely sieved compost

form of protection is given, to prevent cats from using this territory as a toilet, which will surely be disastrous. When the seedlings are large enough to handle, thin them out, leaving the strongest one at each station.

Carrots should never be allowed to dry out or they will split; give a light shade in very strong sunlight to moderate the temperature. During the later period, when growth is developing rapidly, secondary growths may appear around the edges of the crown. These must be removed as soon as they appear. Apply peat or some similar material between the rows to prevent greening of the crowns, which would render them useless for showing.

Harvesting
The following method is the ideal way to lift long carrots. Scoop out the compost immediately around the crown to create a saucer-like depression, then, keep filling this up with water from a hosepipe or watering can. When you think that the water has penetrated well down, gather all the foliage in one hand and start to draw out the carrot, continuing to add the water all the time. As the carrot is drawn up, the water will move down more easily, preserving that fine 'whip', which so enhances your specimen's appearance.

Show preparation
This task requires much patience. The specimen should never be allowed to dry, so, as soon as it is pulled, cut off the tops to about 4in (10cm), lay the carrot in a flat trough of water and cover it. When you have lifted the desired number, start to wash them. This must always be done with a sponge (not a brush), and soapy water, which must be rinsed off finally with clear water. The movement should be round the carrot, not along its length, except at the whip end where care is needed if you are not to break it off. After washing, wrap the carrots in clean white paper (not newsprint), place them in a long box, cover them with damp paper or cloth, and store in a dark shed.

Selection
When choosing your final exhibits, remember that the body of the carrots must be approximately the same thickness and length, the colour must be the same intensity of red and each specimen must be smooth and free from tiny rootlets. The whips should be as long as possible. Matching long carrots is one of the most difficult tasks in exhibiting.

Staging
Display these carrots in a fan shape, with the whip ends towards you. Spray with water and cover them with a wet cloth to keep them looking fresh (remove just before judging starts).

Varieties and points
Many of the long-carrot strains exhibited today are perpetuated by the exhibitors themselves, who have spent many years re-selecting. Good varieties available from commercial seed-houses are St Valery and New Red Intermediate.

Points awarded are: condition 5, form 3, colour 6 and uniformity 6, totalling 20.

Carrots, Stump-rooted

These do not require the extensive preparation needed for long carrots, although, before sowing, the compost or border must be free from stones and other debris, which may scar the immature roots. It is a simple matter, even in the open border, to make holes approximately 8in (20cm) deep, 3in (7.5cm) apart (using a dibber), and fill them with a special compost, as for long carrots. Again, it is advisable to sow three or four seeds per station, and thin out as it becomes necessary.

Harvesting
Water well before trying to pull the carrots, because it is advantageous to keep a straight tap root on the end.

Show preparation
Stump-rooted carrots require the same preparations as long carrots.

Selection
Choose the carrots which have a definite stump before the tap root, rather than a gradual tapering. When boxing for transporting, wrap them in a damp cloth or paper to retain freshness. The RHS Handbook does not state the virtues of good stump-rooted carrots, but the best exhibits are approximately 6in (15cm) from crown to stump.

Many firms sell good stump-rooted carrots, the main variety being Chantenay Red Cored. Observe the types preferred by judges to gain experience (the RHS does not give a points' system).

Cauliflowers

The only way to be sure of having good cauliflowers at any given time is to plant large quantities of varieties that will mature in succession. They must produce plenty of leaf to cover a dome-shaped curd, which must be of even texture throughout, with no signs of lumpiness, and perfectly white, with no greening where light has penetrated.

Selection
With experience, you will be able to choose your specimens two or three days before the curd is fully developed, and tie the leaves neatly round the curd to prevent light entering. (This explains the need for plenty of foliage.) At this stage, examine the curd to make sure there are no foreign bodies present, eg caterpillars.

Also, it helps to place an easily seen marker by each specimen, as this will save an enormous amount of time when cutting. It is wise to mark ten per cent more than you need because there are always

Cauliflowers staged for judging

some which do not come up to the required standard on final examination.

After cutting and final selection, retie the leaves loosely to give protection in transit. Large crates, or cases, with plenty of ventilation are essential to prevent 'sweating' or heating up, which will make the curd open.

Tie cauliflower leaves round the curd to protect it from light

Staging

If the curds are required for collection boards (triangular boards with groups of nails knocked through from the back), remove the foliage completely and push the curds onto the nails. Later, surround them with masses of fresh moss-curled parsley. If, however, the specimens are to be exhibited in a specifically defined class, trim the leaves level with the curds at the last moment, and stage according to the schedule.

A light spray of clean water just before judging will give an added look of freshness, which is important, as condition is a major factor in judging cauliflowers. Points awarded are: condition 10, solidity 6 and uniformity 4, totalling 20.

Varieties

Plant a selection of varieties to fill your requirements over the whole season: start with Dominant, Dok Elgon, Nevada, Cervina, foliowed by the Australian varieties of Barrier Reef, Canberra and Mill Reef.

Celery

The natural habitat of celery is by the side of a stream, where water is constantly available but never stagnant. This explains why it should never be allowed to dry out.

Propagation

For the early shows, sow in early February, with successional sowings at fortnightly intervals (until the beginning of April) to ensure a supply through the show season. Celery seed will germinate better uncovered, on top of the soil, as it benefits from being subjected to light. This is one vegetable which must never have a check at any stage. It requires a temperature of approximately 60°F (15°C) and is best sown in a peaty compost in small pots, three to four seeds per pot, and thinned when large enough to handle. The multi-type polystyrene seed-trays are ideal as the plants can be pushed out without disturbance when the time comes to pot them on into 5in (13cm) pots.

Feeding

Start feeding in early April, at weekly intervals, using a fertiliser with a high nitrogen content, and as the season progresses turn gradually to high potash. The site should be well manured the previous autumn.

Planting

Hardening off can start in early May, so that the plants will be ready for planting out in late May to early June. Plant 18in (45cm) apart with the rows 3ft (1m) apart.

Blanching

Start blanching about seven weeks before the date of the show, by wrapping 6in (15cm) wide strips of dark paper round each plant. Repeat this at twelve to fourteen day intervals, increasing the width of the paper by 6in (15cm) at each wrapping, and inspecting thoroughly for slugs at the same time. Unless there is a constant supply of slug pellets round each plant, damage is likely to occur.

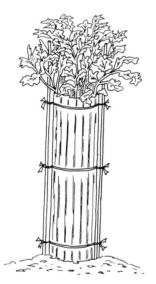

Blanching:
celery in its final wrapping

As each wrapping is made (not too tightly) the stems must be kept straight and not allowed to twist. Retain as much as possible of the productive foliage – it should be long, green and fresh (no yellowing). Remove any leaves which are of no further use to the plant, as well as any laterals which may appear from the base. To protect against strong winds, a couple of stout canes can be inserted, one on each side.

Harvesting

Lifting must be delayed as long as possible to keep the sticks looking fresh and crisp. Trim off the roots to leave a firm base and remove any undesirable stems until there is a perfectly formed, unblemished stick remaining. As each specimen is trimmed and washed it should be wrapped in clean white paper.

Before packing and transporting to the show, it is wise to tie a string loosely round the foliage to prevent the stems getting broken, as they are very brittle. Most exhibitors of celery have long boxes (specially made for transporting), lined with foam rubber and well ventilated to prevent sweating.

Staging

Celery is usually laid flat on the bench, and should be checked at the last minute before staging for any signs of heart rot, which can occur overnight. Trim off a thin slice of the base to show a fresh cut. Every vegetable has a 'back' and 'front', and there will be one side of your celery which shows the heart to best advantage – make sure it is uppermost. Finally, cover with a thick damp cloth to retain that fresh, glistening appearance.

Occasionally the schedule will specify that the celery has to be a certain colour, so make sure you have the right variety.

Varieties and points

The most popular varieties on the showbench are Giant White, Ideal (pink) and Giant Red.

The points awarded are: condition 7, size 4, solidity 5 and uniformity 4, totalling 20 maximum (which makes celery a 'must' for a collection).

Celery, Self-blanching

This type of celery (stouter and shorter) is not as popular on the showbench as the blanched varieties. Propagation is the same as for blanched varieties, but it can be delayed until about mid March. Staging is the same as for blanched varieties.

Planting

When hardened off (by early June), plant out in blocks or beds, six or seven plant wide, with 10in (25cm) space either side. It is not strictly correct to describe this celery as self-blanching, because it depends on this close planting to produce those creamy-white sticks. This block method makes it necessary to provide some protection against the light – a screen of black polythene 20in (50cm) deep, round the sides will be sufficient.

Harvesting
When lifting this type of celery, take all the sticks as you come to them, even though some may be wasted for show work. Picking at random would cause the surrounding sticks to become green.

Varieties
Varieties most widely used are Avon Pearl and Lathom Self-Blanching. This subject is not listed in the RHS Handbook so it is for the observant showman to find out by experience what is required!

Cucumbers

Only greenhouse cucumbers are grown for exhibition – those which produce male and female flowers, and those which produce female flowers only (usually referred to as Fl hybrids). The former type requires far more attention than the all-female type, because pollination must be prevented, so all the male flowers must be removed as early as possible.

Propagation and growing methods
Cucumbers require a fairly high temperature, approximately 65°F (18°C), and a very humid atmosphere. Seeds are best sown in small pots or soil blocks in a loamless or John Innes No 1 compost. As soon as the seedlings are fully developed, move them into 5in (13cm) pots and do not allow them to dry out.

They grow quite well in gro-bags and should be planted out when the second pair of true leaves are developing. Remove all laterals to a height of 24in (60cm) or the fruits will rest on the ground. Let the main stem grow straight up, but stop all laterals at the third leaf. Some varieties may be allowed to produce fruits on the main stem as well as the laterals, but care must be taken not to overcrop or male flowers may be produced (eg Femspot). Show specimens must hang free, or they will be marked by contact with the hairy stems.

On cucumber plants, stop all
laterals after the third leaf

Selection
Cucumbers are usually shown in pairs, so it is essential that they are
uniform in all respects. They must be fresh, straight, with short
handles, and the flowers still attached. It may be necessary to cut
one a little earlier than the other so as to make a matching pair. As
soon as they are picked, wrap them in polythene (leaving both ends
open) and store in a cool place.

Varieties and points
Good varieties for exhibition are Telegraph and Conqueror, which
are open pollinated, and the Fl hybrids Pepinex 69 and Virgo A.
    The points awarded are: condition 10, form 4 and uniformity 4,
totalling 18.

Leeks

Leeks are usually grown from vegetative growths which occur on
the flower head along with the seed-pods. When these growths are
very few, they may be encouraged by cutting away some of
the seed-pods in August. There are two types of vegetative
development to be found on seed-heads: those which form small
bulbils or pips (about the size of a small pea), and those which
produce growths like grass seedlings (about the size of a pine
needle). Both are capable of producing first-class show specimens.
Keep the seed-heads bearing the growths in a healthy condition, so
that propagation may begin in December or January. Discard any

extra large growths, which have a tendency to bolt the following July.

If care is taken, these vegetative growths will come away freely from the seed-head and can be pricked off into seed-trays containing a reputable seed compost. They will soon develop roots of their own, although they do not need a high temperature; about 55°F (13°C) is ideal until roots have developed, then lower to 50°F (10°C) during the day and 45°F (8°C) at night. Do not push them into growth too early or bolting may occur.

The vegetative growths on the flowerheads of leeks are used for propagation

By early March, they will be growing steadily, so they can be potted on into 5in (13cm) pots, in a loamless or JIP2 compost. As the flags (leaves) get longer, insert an 18in (45cm) split cane, with a tie, to support them in an upright position. By this time they should be sturdy plants, and a further move in early May into 7in (18cm) pots will carry them on until planting time, which is roughly mid June. An application of a reputable feed can be given before planting out in the border, 16in (40cm) apart, with 24in (60cm) between the rows.

Border preparation
Before planting out, give the border a liberal application of well-rotted manure, incorporating it between the top and second spit down. The lower spit is best kept at the bottom, but must be well

Leek plant in 5in (13.5cm) pot
ready for hardening off

Blanching leeks in a
bottomless pot and raised
field tile (*below left*);
(*right*) Leek being blanched
in a piece of plastic drainpipe

broken up to enable the roots to penetrate more easily. There are testing kits on the market which will help you decide how much lime your soil needs, which fertiliser to apply and in what quantities. There is one golden rule to observe: never apply lime and fertiliser at the same time. It is usual to apply the lime, if required, then the fertiliser when forking over (or rotovating) a few days before planting, so as to allow the soil to settle.

Planting

It is wise to knock a plant or two out of the pots first to make sure that enough roots have been produced. If the plants are given a thorough watering before planting they will leave the pots quite easily without any loss of roots. Those ties to the canes, which were made earlier on, will be of enormous benefit at this time, minimising the damage to the flags as they are removed from the pots.

Planting out is a task that requires a lot of time and patience if you are to make a good job of it. The best way is to mark out the rows running east to west, and when planting see that the flags are all pointing along the direction of the row; this enables the plants to receive maximum light.

The depth of planting is important and should be about 2in (5cm) below the general level of the border. At this time of the year, the ground is usually quite damp, so watering-in should not be necessary; the roots will soon penetrate the border.

Blanching

After about a fortnight the plants will have become sufficiently established to start blanching. There are several methods used, and three which are worthy of recommendation. Firstly, place an ordinary 3½in (8cm) field tile, 12in (30cm) long, over the plant, allowing it to peep out at the top. Or cut the bottoms out of old 8in (20cm) pots and place them over the plants followed by the field tile: as the leek extends, the tile can be raised and 4in (10cm) of peat put in the pots, allowing the tiles to be lowered on to the peat. Or put a 6in (15cm) field tile over the plant, then when the leek is well developed put a 4in (10cm) plastic drainpipe off-cut, 16in (40cm) long, inside the field tile and push it down to the bottom. This will also give added support in strong winds. Stout canes should be inserted on the outside of the tile in exposed positions. Any of these methods can produce first-class show specimens.

29

Black polythene discs, with 2½in (6cm) holes in the centre, placed over the leeks and resting on the tiles, will prevent light entering and greening the shaft of the leeks.

Feeding
Leeks are lush growers and heavy feeders, but take care when applying nitrogen, or splitting will occur. The safest way is to start feeding with equal parts of nitrogen, phosphate and potash, gradually raising the latter until a ratio of three parts potash to one part nitrogen is reached. This will help to mature the plant and reduce the amount of damage by splitting. Feeding can start in early July and proceed at ten-day intervals until the end of August. Some growers rely on manuring the previous autumn and adding plenty of base fertiliser in spring.

Harvesting
When you have a few minutes to spare, it is worth selecting some plants of approximately the same thickness and marking them. This will make selection easier when you start to lift. Tie up the flags, clear of the ground, to prevent them splitting when they are taken out of the tile.

Show preparation
Tease as much soil as possible from the roots while the plants are on the plot. Draw the tiles off over the root end (not over the flags), then compare their uniformity. Strip off the dead or damaged flags first, and hope that the end product is a clean white shaft! Sometimes, there are brown marks on the inside of the flags; this means taking off another one which reduces the thickness of the shaft. A good blanched leek will be white up to the 'fast button' (see Glossary) and not beyond. As each leek is washed, tie a green string at the top of the blanch to prevent further splitting and show off the uniformity of length.

Staging
Leeks are usually shown in sixes or threes, and the best way of displaying them is with their roots (all clean and white) pointing towards the judge. The flags should be neatly placed and, in particular, free from rust. Staging on black velvet gives a good contrast to the white shafts of the leeks, but check first that this is permitted.

Points and qualities
Leeks are awarded maximum points of 20, and so, naturally, are to be seen on every collection of six or more vegetables. The points awarded are: condition 8, solidity 8 and uniformity 4. Judges look for long, white, straight, thick shafts, with no tendency to bulbing. It is important that the blanching should not go beyond the 'fast button'.

Perpetuation of strains
Leeks are prone to certain virus diseases, so, in order to keep a good clean stock, showmen raise a few from seed periodically, though the resultant specimens are very variable and only a few are suitable for further propagation.

Varieties
The regular prizewinners are usually labelled 'own strain' because most exhibition leeks are raised by exhibitors who keep their own strain. A favourite of the listed varieties is Prizetaker.

Pot leek in 6in (15cm) plastic pot ready for hardening off (see overleaf)

Leeks, Pot

These are raised like the blanched type, but planting outdoors is often a little later. The important thing is to find a strain in which the flags are very compact, so that, when an outer skin has to be removed, it does not unduly increase the length of blanch. This crop is extensively grown in the North East where prizes of great value are at stake.

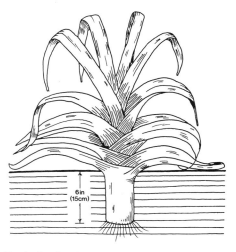

6in
(15cm)

Pot leeks should be earthed up to 6in (15cm)

Culture
The site for pot leeks usually consists of a raised bed supported by planks to enable the grower to attend plants individually. Manure in the autumn, with a base dressing in the spring. Continue feeding at intervals of ten to twelve days, using a proprietary brand of top dressing. The technique of blanching pot leeks is vastly different from the blanched type, because great care must be taken to see that the specified limit of the blanch, which is stipulated in the schedule, is not exceeded. In recent years, a different type of pot leek has

(*top left*) Pot leeks (TL strain)
(*top right*) Collection of vegetables: celery (Ideal), leeks (Prizetaker), carrots (New Red Intermediate), onions (Kelsae) and parsnips (Tender and True)
(*below*) Five potatoes (Pentland Ivory)

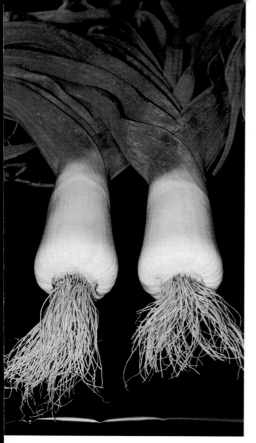

emerged: instead of the shaft being the same thickness all the way down, it is wedge-shaped, the top of the blanch being narrower than the bottom, but the graduation must be even with no sign of bulbiness (see page 33). Blanching is done gradually, and top leek growers check regularly that the length of blanch is not being exceeded.

Staging
Pot leeks are usually shown in pairs or threes. The principles of presentation are the same as for the blanch type.

Varieties
Again, it is not possible to give varieties a specific name, because most exhibitors have their own strains, setting enormous value on them and always striving for improvement.

Points
The points system is different but still totals 20: condition 7, size 7 (cubic capacity of blanched shaft), solidity 5 and uniformity 4. The cubic capacity of the blanch used to be the main criteria, but quality is now given increased consideration.

Marrows

In the exhibitor's world this vegetable does not rank very highly, as it is awarded only 10 points (condition 6, uniformity 4). The great specimens seen in the heaviest section are not to be compared with the tender young fruits in the marrow classes, which may be in pairs, or sometimes threes.

Producing three identical marrows requires more than a little skill. The best way is to gather them all from one plant, for, even if you had twenty plants, you would be unlikely to find two producing the same type of fruit.

Propagation
Sow seeds singly in small pots, in May, under glass, then pot on ready for planting out in mid June, as soon as all danger from frost has passed.

(*above*) Six large onions (Kelsae)
(*below*) Stump-rooted carrots (Favourite)

Culture
Plant on a mound, under which is a plentiful supply of manure. Let only one leader develop until a good strong plant is produced, then allow the fruits to set in succession. If few insects are present, pollination may have to be done by hand. The object is to get three successive fruits growing at the same time on the same plant.

Raise marrows on glass to admit light to the underside

Marrows must be young and tender when shown, so when the first selected fruit on a plant is about 12in (30cm) long (the ideal length), cut it off and store in a cool, dark shed. When the next one reaches 12in (30cm), cut it and place it with the first. Repeat with the third. This is certainly the best way to grow three marrows which will match in all respects. While the fruits are growing, raise them slightly on small sheets of glass, so that the undersides will match the colour of the rest of the fruit.

Staging
They are usually laid out on the flat bench, side by side, with the flower end towards the front.

Variety
The most attractive variety for this class is Bush Green.

Onions

This is probably the most prominent of all exhibition vegetables. From the rawest novices to the most accomplished professionals, countless exhibitors centre their pride on growing the best onions. Today very valuable prizes are offered for the best set of dressed onions, with a lot of interest focussed on the heaviest specimen. Without doubt, to grow the heaviest onion you need a bit of luck!

Propagation

Christmas Day is renowned as the ideal time for sowing onion seed and probably the wives are responsible to some extent for this (traditionally, perhaps, to get the men out of the way while they cooked the Christmas dinner). Be that as it may, there is no doubt that the onion requires a long season of growth and, if the task is not carried out on 25 December, it should not be delayed much longer.

Sow seed thinly in small pots and place in the greenhouse in a temperature of 60°F (15°C) for a few days, until germination has taken place. Then lower the temperature to 50°F (10°C), which is quite satisfactory for producing good sturdy seedlings. As soon as the seedlings are growing out of the loop stage (see overleaf), they are ready for pricking off into 3in (7.5cm) pots, using a John Innes seed compost or one of the more widely used peat composts. The reason for pricking off so early is to prevent any damage to the long initial root. These seedlings can easily be teased out of the compost at this stage, so reducing any slight check which may occur. After potting, always place them as near to the source of light as possible, for they are really long-day plants, and you should aim to have large, sturdy, healthy plants for planting out at the end of April or early May.

The sun will raise the temperature in the greenhouse during the day but, of course, by late March, some ventilation will probably be necessary. The occasional feed may be required by the middle of March, but only at the strength recommended by the manufacturer. By the end of the third week in March the seedlings will need potting on into 5in (13cm) pots using a reputable loamless or JIP3 potting compost. Water the day before, so that the plants knock out freely from the pots. When potting, the compost should not be rammed down hard – just give the pots a couple of taps on the bench, and leave sufficient space at the top for watering.

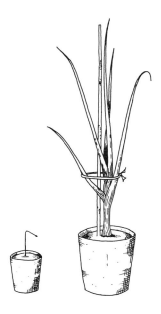

(*left*) Onion seedling at the loop stage in a 3in (7.5cm) pot; (*right*) seedling potted on into a 5in (13cm) pot

Staking

As the plants become established in the larger pots, give them some support by inserting an 18in (45cm) split cane, about 1in (2·5cm) from the plant, and tie about halfway up. A further tie higher up may be necessary before hardening off, to prevent any of the leaves getting bent or broken.

Hardening off will, to some extent, be governed by the weather, but usually late April is about the right time to put them into a cold frame or cold greenhouse. Feed before planting out, to prevent exhaustion of nutrients.

Site preparation

The onion bed should have been well manured the previous autumn, so that tests can be made for lime and nutrients as soon as the weather permits in the spring. Make the lime test first, and apply the recommended amount, if required. Follow with a test for nutrients, which should be worked into the top 8–9in (20–22cm), a few days before planting.

Planting

Unless some protection can be given around the border, early May is quite soon enough to plant out, and, even then, a screen of muslin or one of the recommended wind breaks should be used.

Do not place a solid fence around, as a half-break screen is more satisfactory.

By this time, the leaves will be 2ft (60cm) tall, but don't rush to get them out, wait until the weather seems settled. Water the pots the day before planting out, so that no watering will be needed unless the border is dry.

Before starting to plant out, mark out the rows, so that you know exactly how many you can get into the space permitted. When planting out, check that the leaves are all pointing the same way and that the cane is on the north side so as not to deprive the plant of any light. Many onion beds today are given some kind of temporary cover, which, apart from protecting against wind, will also raise the temperature round the plants. Remove the cover by about the middle of June, or when the weather is favourable. The ideal temperature for optimum bulb development is 68°–75°F (20°–23°C), with a day length of 15½ hours or more, and a high light intensity. In Scotland and the north of England, some exhibitors are now leaving the covers on right through the summer.

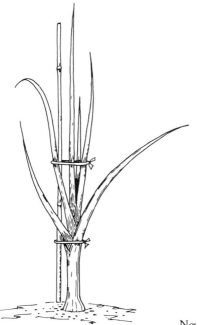

Newly planted onion

When the plants have become established in their permanent quarters, cut the ties and remove the canes. No doubt, a few leaves will bend over, but this cannot be avoided, and the plants will soon adjust to the new conditions, supporting themselves (wind permitting). The main task then is to keep the hoe going, so that there is no competition from weeds.

Feeding

Feed on the bed at your discretion, but go steady with the nitrogen. Start with a fertiliser of the ratio 2 parts nitrogen, 1 part phosphates and 1 part potash; as the season progresses, change to 1 part nitrogen, 1 part phosphates and 3 parts potash. This will produce a solid, ripe bulb, with good keeping qualities.

Summer treatment

Towards the end of July, when the outer skins start to split and begin to look untidy, take off the unsightly skins along with any dead foliage. Take care not to pull down on the leaves, or damage may be done to a sound skin, which could be the one needed for final ripening. This operation allows the sun to get at the newly exposed skins, which helps considerably towards the production of a well-conditioned specimen.

Harvesting and ripening

As bulb development continues, keep a watchful eye on the types for matching up, though it is not until the fateful day of lifting that you can be sure. A day or two before the desired specimens are required, ease them up slightly with a fork, to break off some of the roots and avoid lifting while in full growth.

Immediately after the bulbs are pulled, cut off the tops and roots, leaving a neck of about 6in (15cm). Then wash the bulbs in soapy water, taking care not to get any water down the neck. After wiping dry with a clean towel, lay them on the greenhouse bench, in boxes lined with soft clean hay or similar material. Then place them in the sun for a day or so. Do not leave them in the direct sunlight for too long, or a further skin may split, which would ruin them for showing. A suitable place for further ripening is a building which is warm, with plenty of air movement. (It is very noticeable how exhibits will condition while on show in a marquee on a sunny day.)

Show preparation
When you have decided which onions to stage, tie their necks with natural raffia. There are two popular methods: one is to wrap the neck 2/3in (5 or 7·5cm) from the top, tightening as each wrap is made, so that the neck appears as thin as possible; the other is to fold the neck over to make it look thin. The fold should always go to the back of the onion and be neatly tied down.

Transporting
Transporting these fine specimens to the show without damaging them in any way requires skilful handling. Some exhibitors make special boxes, lined with foam rubber. Others wrap the onions well in paper or wood wool to cushion any bumps in transit. I use both methods. Most of all, prize onions must never be lifted by the neck, as this will crack the skin at this point and reduce their chances.

Staging
Onions are usually staged in threes or sixes – to see a set of onions perfectly matched and perfectly staged is a sight to behold! Stage in the form of a triangle, three raised up at the back, then two a little lower, with the very best specimen at the front, where it is shown off to best advantage.

Some showmen have a permanent stand, covered in black velvet to show the onions at their best. I use 4in (10cm) plastic piping, cut in various lengths and covered with a large piece of black velvet. Another method uses a basket large enough to stand five onions round the bottom, with the best one on the top.

Points and qualities
The points awarded for onions are a maximum of 20: condition 5, size 5, form 5 and uniformity 5. Good onions have thin necks, firm bulbs and are well ripened (especially for the later shows). Thick-necked bulbs with broken skins are down-pointed.

Varieties
The most notable varieties for exhibition today originate from Ailsa Craig, which dominated the showbench years ago. The varieties most popular on the showbench are Kelsae, Mammoth and Oakey. The majority of top exhibitors perpetuate their own strains.

Onion Sets

Onion sets are produced from small, late-sown seedlings, which are given a special treatment to arrest the development. Sets should be obtained from a reliable source or they may turn out to be very disappointing, and bolt during the following summer.

Site preparation and planting
Plant as soon as the weather permits after you have bought the sets, whether autumn or spring. The border need not be so heavily manured as for the large bulbs, because the size required for showing is usually 'not to exceed 8oz (240g)'. About 2oz per sq yd (60g per sq m) of bonemeal will be adequate if worked into the top 8in (20cm) of soil.

Plant in five-row beds, about 6in (15cm) apart in the row and the rows 12in (30cm) apart. Not much attention is needed, except to see that the birds do not pull them out before they are rooted. Keep the hoe going whenever conditions are suitable. Spring-planted sets usually mature about fifteen weeks after planting, which is quite a few weeks ahead of bulbs grown from seed (autumn-planted sets are slightly earlier).

Much progress has been made recently in the production of onion sets, and there are now fairly uniform globe specimens as well as the flattish types.

Selection
To gauge when the onions are the right size, measure the diameter of one or two bulbs which weigh 8oz (240g), or just over, when newly pulled. (They will dry out whilst conditioning and will be just under the specified weight.) Pull onions which match this diameter, and remember that they must match in all other respects. Bulbs from sets are usually a rich, attractive colour.

Show preparation
The tops must be tied down neatly, and, under the National Vegetable Society's rules, you must use natural raffia. If a stipulated weight is mentioned in the schedule, then all specimens will be weighed; if one is overweight, the only award your exhibit will receive is an NAS (Not According to Schedule). The RHS Handbook does not classify a points system for onion sets.

Staging

Although these onions are only small, they make an attractive exhibit if displayed in a triangle, like large bulbs, on black velvet. They are often stood on curtain (or plastic) rings, which should be covered with black velvet.

Varieties

Good show varieties are First Early, Fenland Globe or Sturon.

Parsley

Parsley requires rich ground and plenty of space, at least 12in by 12in (30cm by 30cm); or it may be grown round the outside of the plot, for easy picking. Seeds give a very staggered germination and are best pricked off while in the seed-leaf stage, or sown in 'multicell' plant trays (three or four seeds per cell), using a reliable seed compost. Thin out, leaving the strongest seedlings to develop. Start sowing in March, with a further sowing in April. Coldhouse temperatures are adequate at this time of the year, so the plants can be hardened off during May, and planted out at the end of the month.

Culture

Parsley resents dryness at any time, so watch it carefully. If a plant starts to die back through lack of water, it seldom recovers sufficiently to produce a good specimen.

Staging

Parsley is permitted as a garnish for collections – its tightly curled foliage certainly adds to the attraction.

It is allocated 5 points, but must be clean, fresh and free from disease.

Varieties

Good varieties for exhibition are Bravour and Paramount.

Parsnips

Parsnips are grown in more or less the same way as long carrots (see page 17), but bore holes to a depth of about 4ft (120cm), as they require an even greater depth than carrots.

Sow from early March onwards, as they will germinate at fairly low temperatures (though very early sown crops are more susceptible to canker than those sown in April). Sow several seeds at each station (stations should be about 12in (30cm) apart). If you use large barrels, five stations per barrel is ample.

Culture
When seedlings are large enough to handle they must be thinned down, leaving the strongest one. If the weather is dry, copious watering is the order of the day. As the parsnips reach maturity, the broad shoulders start to push the soil upwards, so more peat or compost must be added to prevent them from becoming green (due to exposure to light). This vegetable is very easily damaged, so be careful that the roots are not caught when hoeing. An open wound would probably lead to an attack of canker.

Lifting
When lifting show specimens, follow the same procedure as for long carrots. A lot of water is needed round the roots to penetrate to the lower levels and loosen the roots as they are pulled. Long roots need to be pulled quite hard, so take care that the tops do not break off in the process.

Show preparation
Lifting parsnips is one of the tasks which must be delayed as long as possible before the show, as they so easily discolour. Remove any fine rootlets carefully from the length of the root, and wash with a sponge. When washing, the movement should be round the root (not along its length), except at the whip end. Take care not to break the whip off. The end product must be a long, white, tapering specimen.

Staging
At larger shows, the schedule may require six specimens to be staged, but this is not easy to achieve because the roots are so susceptible to damage. Only 3in (7.5cm) of foliage is permitted and roots are displayed on a background of black velvet to emphasise their whiteness.

Points and qualities
Good parsnips must be long and tapering, with well-developed shoulders, smooth white roots and no blemishes. They are

awarded a possible 20 points: condition 6, size 4, form 4, colour 2 and uniformity 4.

Varieties
The best varieties for exhibition are Tender and True, and Hollowcrown.

Peas

This crop likes a new turf which was turned over the previous autumn, and given an application of lime to give it a pH value of 6.5. The main problem on new land is the possible presence of wireworms (see page 62). Use a low-nitrogen fertiliser, as peas take in their nitrogen from the air, through nodules on the root system.

Cut off the smaller pea seedling at soil level

Propagation
Pea varieties for exhibition are usually tall-growing. They require a growing season of about 15 weeks, if started under glass, or 17 weeks if sown direct into the ground. Most peas for exhibition are sown in small pots in a greenhouse, using a clean, reliable compost. If two seeds are sown per pot, then the stronger one can be retained, cutting off the other at soil level, so as not to disturb the root system. If sowing is delayed until the end of April, a good selection of pods should be available from early August onwards. The plants can be put in their final quarters by late May, when danger from keen frosts has passed.

Training

As the peas are tall-growing, some form of structure will be necessary to support them. An ideal method is to erect strong posts, one at each end of the row, with some intermediates at intervals of 10ft (3m) and 6½ft (2m) high. Stretch a wire along and attach it to each post, at a height of about 6ft (180cm). This will support 7ft (210cm) canes, which are pushed into the ground about 10in (25cm) apart and tied to the wire. The plants are put out, one to each cane, and secured to this with a sweet-pea ring. They are grown on the cordon system (see below) and further rings are added as they grow.

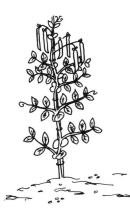

Pea plant secured to a cane with sweet-pea rings

An alternative method is to stretch 6in (15cm) mesh netting, 6ft (180cm) wide, along the row and attach the peas to the netting with rings. It is best to trim off the tendrils and weave each plant up the net as it grows. The peas will sway with the net when the wind blows, so reducing the amount of damage.

Culture

Lateral growths will spring up from the base of each plant and from each leaf axil. These must be removed so that all the energy is directed into the main stem. These laterals will continue to appear until the flower stems form. There is then a continuous supply of flowers, produced from each consecutive leaf axil (this is known as the cordon system). In order to obtain a supply of pods at the right time, you must reckon that approximately one month is needed from flower-bud stage to maturity. This varies of course according to climatic conditions – a day or two either way. Any flower buds which appear before this time must be removed.

Selection
Before deciding which pods to retain for show, hold them towards the sun and count the number of rudimentary fruits in each pod. Any that have less than eleven fruits, deformities, or missing peas, must be discarded – do not allow them to remain on the plant. Some varieties produce their pods in pairs, so that by retaining three sets, you have a succession of six pods per plant to choose from. If, however, they are produced singly, you must allow another two or three to develop before taking out the growing point, two leaves beyond the last pod.

Harvesting and Transporting
When harvesting, it is a good idea to use a large flat box or basket lined with a few carrot tops. Using a pair of scissors, cut the pods off, with about 1½in (3·5cm) of flower stalk attached. Do not touch the pod, or the bloom will be rubbed off; always handle by the stalk! If they are laid gently on the carrot tops, all pointing in the same direction, then little or no damage will be done. A final check against the light for any deformities, of any kind, is always advisable. To stand any chance of winning in top competition, each pod must contain twelve or thirteen fruits, all evenly matched.

As you check them, pack them in a suitable container, lined with carrot tops (fronds only). Provided you don't exceed three layers, with the carrot tops between each, the pods will remain fresh and the bloom will be retained.

Staging
The number of pods required for an exhibit varies from six for local shows, to fifteen for the larger ones. Because there is a succession of five or six pods from each plant, it is possible to cover three weeks or so from one sowing. At championship shows, special boards are provided, covered with black cloth. All the pods are arranged at the same level, all pointing the same way, with a slight space between each. Some shows provide plates, and the pods are then arranged like a cartwheel, with all the stalk ends forming the axle.

Points and qualities
Peas are awarded a maximum of 20 points: condition 8, size 4, fullness of pods 4 and uniformity 4. Pods must be large, fresh, with bloom intact, and full of tender peas.

Varieties
Suitable varieties are Show Perfection and Achievement.

Potatoes

Although potatoes are the most widely grown of all vegetables, they are also one of the most difficult subjects to match up. As they are propagated vegetatively, if the stock carries any virus disease, it will be handed down to the progeny. Some of these viruses are transferred by greenfly, so potatoes for seed are grown in areas at high altitudes, where greenfly are rarely seen. Certified seed is well worth its extra cost. Potatoes for seed purposes are usually a little larger than a hen's egg.

Chitting
When the seed potatoes arrive, set them out in boxes on the greenhouse bench, where they can get plenty of light, but are safe from frost. They should be set with the rose end uppermost. Several shoots will be produced, but, before planting, these should be restricted to one or two; otherwise the crop will consist of many small tubers, which will be no use for exhibition.

Seed potatoes set up for chitting

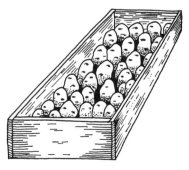

Planting
Potatoes are usually grown on a site where brassicas grew the previous year. They must be given generous spacing when planting. This can begin in April, so that the severe frosts will

probably have passed before they emerge. Place the tubers in well-manured trenches, 2ft (60cm) apart in the row, with 3ft (90cm) between the rows. The manure will help to retain moisture, which is so important in keeping the crop clean and free from scab. It is a tradition that when planting potatoes, you should have a dirty finger and thumb – from pressing the tubers firmly into the manure.

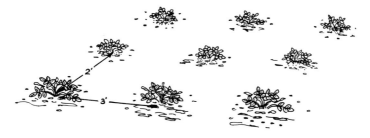

Potatoes emerging and ready for earthing up

A covering of peat directly over the tubers will be beneficial, before applying a good general fertiliser at 2oz per sq yd (66g per sq m), finishing off with soil from the next trench. No lime must be applied as this tends to induce scab.

Precautions must be taken against any late frosts by earthing-up as the growth emerges, or alternatively, by giving a covering of old pea-sticks.

As growth continues, it is well worth while keeping the tops growing in an upright position by supporting them with string, running along each side and held by stout canes. This may seem unnecessary, but if they can be kept growing well clear of the ground it is easier to control pests and disease.

Tubers start swelling when the plant is in flower and, from this this time onwards, strict attention must be paid to watering.

Harvesting
Before starting to lift the crop, you must have sufficient containers, especially if you wish to keep a few tubers for seed, because these must come from good roots only. When lifting, take care to get behind and under the root with the fork – otherwise, sure enough, you will spear your best specimen. The container for show specimens should be partially filled with water, which as the potatoes are put gently in, will prevent any bruising.

Plates of exhibition potatoes, showing their shallow eyes

Show preparation

Never let show tubers dry before washing, as it is more difficult to remove the dirt once it has dried, and that may lead to damaged skins. Soft soapy water is ideal for washing; always use a sponge, never a brush. When washing is completed, mop the tubers dry with a clean towel. Do not wipe, or skin will be knocked off and points lost.

Selection

The specimens must be typical of the variety, with few eyes, and the eyes must be shallow. The tubers must be approximately 4in (10cm) in length for the long types or 2¾in (7cm) in diameter for round ones. Lay them on a table and grade for size, picking out the best matching set. Then wrap in soft tissue paper, finally covering with a dark cloth, to keep out the light.

For transporting to shows a small box may be used. Each potato should be wrapped in an additional paper, and as each layer is

Runner beans (Streamline) (*Harry Smith Collection*)

50

completed place a sheet of bubble plastic over it, before adding a second or third tier.

Staging
The exhibits are usually staged on paper plates, often provided by the show committee. The best way to display potatoes is to arrange them in a circle, with the rose end pointing outwards and up-wards. Experience will soon tell you how to display them to their best advantage. The exhibit should finally be covered with a damp cloth, to keep it in a fresh condition and prevent greening before being judged.

Potatoes are awarded a maximum of 20 points: condition 4, size 4, shape 4, eyes 4 and uniformity 4.

Varieties
Good varieties for exhibition are: (long coloured) Catriona, Vanessa and King Edward; (long white) Pentland Dell, Dr McIntosh and Mona Lisa; (round coloured) Ulster Classic, Cara and Romano; (round white) Arran Comrade, Pentland Ivory and Croft.

Shallots

As shallots are closely related to the onion family, they have similar soil requirements (see page 38). They are extremely hardy, so can be planted out as soon as the weather permits. There is a traditional saying that shallots should be planted on the shortest day and lifted on the longest. This is taking it to extremes, as it is difficult to get the soil ready in December, but they should be planted out as early as possible.

The ideal size gives about sixteen to twenty sets to the pound.

Planting
Sets should be half covered when planting, using a trowel. Do not push them into the ground, or they will lift as rooting takes place. If larger sets are used, they can be planted in small pots and started

(*top*) Mixed vegetables at the RHS Show (*Harry Smith Collection*)
(*below left*) Swede (Best of All) (*Harry Smith Collection*)
(*below right*) Marrows (Bush Green) (*Harry Smith Collection*)

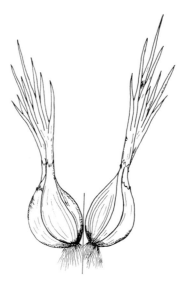

Shallot divided to make two clones

under glass in January, without heat. By the middle of March, they will have made about 3in (7·5cm) of growth; so choose a warm day, rake the border, and make the final preparations for planting. These bulbs will have four or five shoots on each. Examine them carefully, then split them, using a sharp knife, without damaging the shoots. Each section will now have two or three shoots, along with half the root system. This method not only produces larger shallots but, because there are only two or three divisions per clone, they do not become misshapen; so you are likely to get a good number of potential show specimens. The cut edges may be dusted with sulphur, to keep out pests and disease.

Summer treatment consists of keeping them free from weeds.

Lifting
By early July, they should be quite plump bulbs, with good shape. There is a danger at this time that, if left in the ground and subjected to heavy rain, the bulbs will make secondary growth and split. It is wise just to put the fork under them and gently ease them up, snapping a few roots. This will help to prevent splitting and assist in ripening. A few days later, when warm and sunny, they should be harvested and, if possible, suspended in an open shed, to dry off thoroughly and ripen. When sufficiently ripe, they should be trimmed down to one skin, with tops and roots removed.

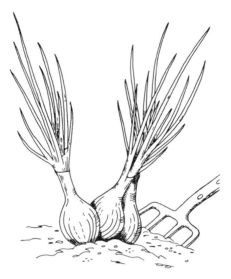

Easing shallots up with a fork, a few days before lifting

Show preparation
To prepare shallots for show, tie them neatly by wrapping round the neck to make an attractive finish, or doubling it over, keeping the best side to the front.

Staging
Shallots are usually shown in sixes or eights at the smaller shows; but at the larger ones the number may be twelve or even fifteen, which makes it considerably more difficult to match up. They are staged on plates containing dry sand, or similar material, and set out in a symmetrical design, so as to catch the judge's eye.

Points and qualities
Bulbs should be firm, round and with thin necks. The points awarded are not very high: condition 5, size 3 and uniformity 4, giving a total of 12. Where a class is for pickling shallots, the size must not exceed 1in (2·5cm) in diameter.

Varieties
The chief varieties grown for exhibition are Hative de Niort, Aristocrat and Giant Yellow.

Tomatoes

To produce fruit suitable for the majority of summer shows, the best time to sow is early March. Seeds can be sown in trays, in a loamless seed compost. They need a temperature of 65°F (18°C) during the day and 56°C (13°C) at night.

Propagation

While the seedlings are in the cotyledon stage, pot them up into 4½in (12cm) pots, using a suitable potting compost. Maintain the same temperatures. In order to produce good healthy plants, feeding must start after about four weeks, with a well-balanced fertiliser. By early May, they should be ready for planting, with the first truss just forming.

Because of the difficulties in maintaining clean glasshouse soil, it is fashionable to use gro-bags. If the instructions are closely followed, it is quite possible to grow first-class plants, with high-quality fruit. The bags should be laid out so that each plant gets its fair share of light. As the plants grow, with the fruit developing, a higher night temperature of 62°F (17°C) will be extremely beneficial, because it will induce the plant to produce better-shaped fruit, of a higher quality.

By the time show dates come along, there should be a plentiful supply of fruit, giving you ample choice of high-quality tomatoes of a good size. If the plants are growing well and are fed regularly, at strengths recommended by the manufacturers, there should be no need to reduce the number of fruits per truss. This measure should only be necessary where there is a tendency for the fruit to be slightly under size. The ideal size is about six to the pound. They must be fully ripe, but firm, with calyces fresh and green.

Selection

Make a preliminary choice while the fruits are on the plant, so that you can match them up for shape, colour and size. The final selection is made when ample fruit has been picked to choose from. When picking, make sure the calyx is still attached to the fruit, or points will be lost. It is best to wrap the fruits in tissue paper or pack them in cottonwool, with calyces uppermost, so they will not bruise in transit.

(*left*) Tomato plant ready for planting   (*right*) Tomato crop, variety Sonatine

Staging

The tomatoes (usually six or twelve) are generally displayed on a plate, in a symmetrical design, with calyces upwards. Sometimes parsley is permitted for garnishing, but it is wise to check with the steward first.

Points and qualities

Tomatoes should be medium sized, round, ripe but firm, with good colour and calyces attached. There are 18 points awarded: condition 6, colour 4, size and form 4 and uniformity 4.

Varieties

Among the most successful varieties exhibited at present are Sonatine and Alicante.

Turnips and Swedes

These two are members of the same family and are grown and prepared in much the same way.

Sowing

Turnips may be sown at intervals from May to August, while the swede is sown in May or early June. It takes turnips eight or ten weeks to mature, so reckon back from the dates of the shows, and sow accordingly. A firmer root is produced from a moderately heavy soil, manured for the previous crop.

Turnips can be sown in drills 12in (30cm) apart and thinned down to a 6in (15cm) spacing.

Swedes may be grown on similar soil, with the rows 15in (37½cm) apart and thinned down to 12in (30cm) apart, when large enough to handle. They may be left in the ground until required, as they are extremely hardy.

Selection

The ideal size for a turnip is about that of a tennis-ball. They should have clean skins, be solid and have shapely tap roots. Swedes need the same characteristics, except for size – they should be approximately 6in (15cm) in diameter. Although these subjects do not command many points, they should be presented as effectively as possible.

Show preparation

All the soil should be washed off. At the same time, remove any leaves which are serving no useful purpose. Any specimens which show the finger-and-toe disease (the same thing as club root in brassicas) must be avoided.

Staging

The number required is usually three, so the best way of displaying them is to put two at the back, with one in between and slightly forward. The tops are usually shortened, but the ruling in the schedule must be adhered to. Fifteen points are awarded, 6 for condition, 5 for size and solidity, and 4 for uniformity.

Varieties

Suitable turnip varieties are Golden Ball and Snowball. The best varieties of swedes are Purple Top and Green Top.

3 Pests, Diseases and Disorders

Bean Aphis (Blackfly)

Both sexes of bean aphis are black; the females produce live young. The sap of the beans is sucked in by such large numbers of aphids that the plants are weakened and so unable to produce a good crop. Control is given by spraying with nicotine, lindane, phorate or malathion. Sprays do however kill off the ladybirds and hoverflies, which are the natural enemies of the pest. Pinching out the tops will give some measure of control.

Cabbage Rootfly

This pest is ashy grey, looks like a housefly, and measures about ¼in (6mm). It is common to all members of the brassica family, but can be controlled by the use of calomel at planting time, or by spraying with phorate or diazinon.

Cabbage White Butterfly Caterpillars

These caterpillars have green bodies, with dark markings as they develop. The natural enemies of this pest are the hoverflies, but as such large numbers of caterpillars appear during August and September it is necessary to spray with such insecticides as resmethrin or diazinon.

Carrot Fly

The carrot fly is shiny black or dark greenish-black, about ¼in (5mm) long, with a wing span of approximately ½in (13mm). Carrots are usually attacked by this pest in late May or early June, with a second attack in August and September. It is advisable to spray with an insecticide such as diazinon, phorate or HCH mixtures, immediately after thinning, as the flies are attracted by the smell. Another method which some people claim is successful is to interplant at intervals with dwarf marigolds; the smell from these camouflages that of the carrot to some extent.

Celery Fly

This serious pest is about ⅛in long, with a tawny brown body (slightly lighter on the underside) and a wing span of about ½in (13mm). The eggs are laid on the undersides of the leaves during April to June. On hatching, the maggots tunnel into the foliage, causing the characteristic mining in the leaves, which reduces their effectiveness and disfigures them, so making them useless on the showbench. The control used for carrot fly is equally effective with this pest.

Flea Beetle

The flea beetle is very small, about $\frac{1}{10}$in (2mm) long. It can be black, bluish-black or greenish-black in colour, with a broad yellow stripe down each wing case. It is a common pest throughout the brassica family. It attacks young plants, making holes in the seed-leaves, especially during sunny periods in April and May. A seed-dressing containing gamma-HCH and captan will give a fairly good measure of control.

Greenfly

These green insects are often called plant lice. Females may be winged or wingless, and both sexes are capable of giving birth to live young. The presence of this pest on any crop, especially potatoes, cannot be taken too seriously. It is largely responsible for the spread of certain viruses. The greenfly visits infected plants and takes in sap, which spends a short time in the digestive juices of the pest before being transmitted to clean stock, so spreading the virus. This is why certified seed is grown at altitudes of 500ft (153m) or over, where greenfly are not likely to be found. An effective measure of control is to spray regularly with fentro or resmethrin.

Leatherjackets

This pest is common to all plants. It is the larva of the daddy-long-legs, and is very tiresome where lawns or grassland are freshly dug.

A full-grown grub is about 1½in (38mm) long, and an earthy colour; the head can be retracted into the body. It will die quickly if exposed to light, or if the soil dries out. Apply HCH or sevin dust during the winter period.

Onion Fly

Adult flies resemble small houseflies, but they are grey with dark triangular marks on the abdomen. Spring and early summer are the times when attacks are most frequent. The maggots from these broods attack the roots of the onion and also burrow into the base of the bulb. A natural enemy is the rove beetle, as its grubs parasitise the maggots of the onion fly. Dusting with 4 per cent calomel will deter this pest.

Pea Moth

The pea moth is very rarely seen but is very troublesome during its short period of activity. It is the pest responsible for the caterpillars which are sometimes found in pods, eating into the fruit. It is to be found mainly in the mid-season varieties; very early and very late maturing crops usually escape. Control is effected by spraying with HCH, fentro or resmethrin.

Slugs and Snails

These pests probably cause more trouble than any other, as they are capable of reproduction all the year round, although the peak period is in the spring. The most widely used material for controlling slugs and snails is metaldehyde, incorporated with bran in pellet form. Another mini-pelleted formulation is methiocarb. The majority of vegetables and other plants are susceptible to attacks from these pests.

Turnip Fly

This is really a flea beetle (see page 60). It attacks the seed-leaves of the turnips and is most serious in May and June. Spray weekly with HCH, while plants are in the seedling stage.

Wireworms

Wireworms resemble golden-coloured worms ¾in (18mm) long. They can remain in the soil for four or five years before pupating, to become click beetles. Potatoes are most vulnerable to attack from wireworms. Apply bromophos at ½oz per sq yd (20g per sq m) during winter.

Blight on Potatoes

The first symptoms of blight are dark brown spots on the leaves. On the underside of the spots is a delicate white mould. The spread of this fungus disease is largely dependent on certain weather conditions which are known as the 'Beaumont period'. This consists of a forty-eight-hour period, when the temperature does not fall below 50°F (10°C) and the humidity of the air does not fall below 75 per cent. When this happens, warnings are sent out by radio or press, so that the necessary precautions may be taken – spraying with one of the fentin formulations.

Botrytis

The first sign of botrytis is usually to be found where leaves have been removed or there is damage – open wounds are ideal for the entry of botrytis spores. The initial brownish colour is followed by a greyish mould. It is more prevalent in areas which remain damp or wet. Benomyl will give adequate control if applied regularly.

Club root on Brassicas

This is a slime fungus that persists in the soil for several years. Spindle-shaped swellings occur on the roots, which eventually rot and so release thousands of spores to overwinter in the soil. Heavy applications of lime will assist in controlling this disease, but, of course, this would not be good for potatoes as it would induce scab. Four per cent calomel dust will give good control – apply one teaspoonful per dibber hole before planting.

Leaf Mould (Cladosporium)

This shows as greyish white spots on tomato leaves, with a brownish mark on the underside. Usually, it appears on the tomato crop in July, as a result of hot and humid conditions. It can be kept under control by good cultural practices. There are, however, new varieties which are extremely resistant to the various forms of this disease.

Leaf Spot on Celery

This can be recognised by discoloured areas which increase in size until the whole leaf becomes a dirty greenish brown and eventually rots. Commercially produced seed is usually given the hot-water treatment to control this disease, but it can occur when exhibitors save their own strains. This is a risk that exhibitors are prepared to take. Spraying with benomyl or zineb at regular intervals will help to keep the plant clean.

Leek Rust

This appears as bright-orange spots, which eventually spread over the whole leaf. Leek rust has become more troublesome in recent years. The spores liberated from the postules cover the leaves, making them useless to the plant, which will eventually die. It is carried over on vegetative material in the soil. One of the most effective materials for controlling rust seems to be Bayleton.

Mildew

Various forms of this fungus are to be found on different vegetables, which take on a white, mealy appearance. Control is usually effected by the application of combinex or karathane.

White Rot in Onions

This fungus attacks the roots and base of the bulb, producing a white mould, which results in yellowing of the leaves. Four per cent calomel dust will give good control if applied to the base of the plants when planting out.

Blossom End Rot

This is a physiological disorder, due to a temporary shortage of water, which in turn results in a shortage of calcium, especially where other nutrients are high. The area where the flower was becomes dark brown and sunken. It can be remedied by keeping the soil sufficiently moist.

Heart Rot in Celery

With this disorder, the young leaves in the heart of the plant turn brown, and the plant eventually dies. It may be caused by excess of certain nutrients such as potassium, ammonia, magnesium; or a deficiency of calcium (caused by a temporary shortage of water).

Tomato Mosaic Virus

This disease is recognisable by small lighter coloured areas in a mosaic pattern. It is more easily seen on sunless days. It is a virus disease, readily transmitted on workers' hands. It is possibly carried over in the soil, or perhaps the seed coat. Many varieties today are resistant to this disease, due to work done by plant breeders. Such varieties are Cura, Curabel or Sonatine.

Glossary

Activator A material that is capable of inducing and accelerating decomposition.

Anaerobic Without free oxygen.

Bines Stems of climbing plants requiring support.

Bolt Produce a seed stem.

Calyx (plural, calyces) The outermost group of floral parts.

Chit to induce young shoots from tubers.

Compatible Capable of existing together.

Cordon Grown on a single stem.

Cotyledons The seed-leaves of plants.

Fast button Lowest point of outer leaf on leek stem (where leaf forms a 'v').

Friable In a crumbly condition.

Formulations Materials combined to a formula.

Half break A fence allowing some of the wind to pass through.

Interplant Growing one crop between the rows of another.

Legumes Plants bearing seed-pods.

Lifted Dug up.

Nitrifying Oxidising to nitrites or nitrates by bacterial action.

Parasitise To kill by parasites.

Postules The parts of plant rusts which release spores.

Rudimentary Undeveloped forms.

Seed-leaves The first leaves produced by a plant.

Spores Germ cells.

Trace elements Plant nutrients required in very small quantities.

Transmitted Carried over from one plant to another.

'Whip' The long threadlike extremity of a tap root.

Index